Energy 60

风吹来的蛋糕
Cakes From Wind

Gunter Pauli

冈特·鲍利 著

王菁菁 译

学林出版社
www.xuelinpress.com

丛书编委会

主　任：贾　峰

副主任：何家振　郑立明

委　员：牛玲娟　李原原　李曙东　吴建民　彭　勇
　　　　冯　缨　靳增江

丛书出版委员会

主　任：段学俭

副主任：匡志强　张　蓉

成　员：叶　刚　李晓梅　魏　来　徐雅清　田振军
　　　　蔡雩奇

特别感谢以下热心人士对译稿润色工作的支持：

姜竹青　韩　笑　杨　爽　周依奇　于　哲　阳平坚
李雪红　汪　楠　单　威　查振旺　李海红　姚爱静
朱　国　彭　江　于洪英　隋淑光　严　岷

目录

风吹来的蛋糕	4
你知道吗？	22
想一想	26
自己动手！	27
学科知识	28
情感智慧	29
艺术	29
思维拓展	30
动手能力	30
故事灵感来自	31

Contents

Cakes From Wind	4
Did you know?	22
Think about it	26
Do it yourself!	27
Academic Knowledge	28
Emotional Intelligence	29
The Arts	29
Systems: Making the Connections	30
Capacity to Implement	30
This fable is inspired by	31

一天，在岛上的农场里，一只小猪一边嚼着胡萝卜，一边向他的邻居抱怨着。
"为什么人类要给我们吃这么多胡萝卜啊？"他叹了一口气，"我已经厌倦了整天嚼胡萝卜！"

One day, on a farm on an island, a pig is chewing and chewing on a carrot, and complaining to his neighbour.

"Why do they give us so many carrots to eat?" he sighs. "I'm tired of chewing on them!"

一只小猪嚼着胡萝卜

A pig is chewing on a carrot

我们长着很多牙齿

We have lots of teeth

"别抱怨了,好好享用吧!"他的朋友回答说,"这些胡萝卜已经被切成块了。"

"唉,人类只知道我们长着很多牙齿,但是我想他们不知道我们没有耐心把这些都嚼完。"

"Don't complain, enjoy!" says his friend. "These carrots have already been shredded to pieces."

"Well, the humans know we have lots of teeth, but I suppose they don't know we lack the patience to chew through all of them."

"那么,他们为什么要给我们啃那么多胡萝卜啊?"

第一只小猪哼了一声:"因为超市不想要这些胡萝卜呗!"

"为什么?这些胡萝卜怎么了?"第二只小猪问。

"So then why do they overload us with carrots?"

The first pig snorts. "Because the supermarkets don't want them!"

"Why? What's wrong with them?" the second pig asks.

超市不想要它们

The supermarkets don't want them

只想要长得笔直的胡萝卜

Only want carrots that are straight

"就因为这些胡萝卜长弯了！"
"可还是很好吃啊！"
"那当然，但是超市的顾客们看了太多电影，他们只想要那些长得笔直、长成完美圆锥形的胡萝卜。"

"Only that they're crooked!"

"But still tasty!"

"Of course, but buyers at supermarkets have watched too many movies and they only want carrots that are straight and perfectly conical."

"太可笑了。"第二只小猪摇摇头,"也许这些人应该改去做胡萝卜蛋糕。"

"是呀,他们应该收割所有胡萝卜,把它们储存起来,这样一年到头就有的吃啦!"

"Ridiculous." The second pig shakes his head. "Perhaps these people should make carrot cake instead."

"Yeah, they should harvest all carrots, store them somewhere, and have a supply all year round."

这些人应该改去做胡萝卜蛋糕

People should make carrot cake instead

鲜榨胡萝卜汁

Freshly squeezed carrot juice

"没错,很有道理。我就很喜欢胡萝卜。吃胡萝卜让我看起来漂亮又粉嫩。"

"人类应该把那些块头大的胡萝卜榨成胡萝卜汁,每天都喝鲜榨的,然后把那些被榨完的碎末喂给我们吃。"

"Yes, that makes a lot of sense. I like my carrots. And they make me look nice and pink."

"They could turn the biggest carrots into juice, freshly squeezed daily, and leave the nicely shredded leftovers for us."

"你的想法很高明，可是，在这座岛上储存胡萝卜需要很多能源——夏天要保持凉爽，冬天要防止冰冻。"

"我们是住在岛上的，对吧？"

"是呀，没错。那又怎样呢？"

"是这样的。在白天，地面变热，可海水能够保持低温；而在夜间，地面变凉，可海水又是热的了。"

"You are smart, but it'll require a lot of energy to store carrots on this island – to keep them cool all summer and stop them from freezing in winter."

"We live on an island, right?"

"Yes, we do. So?"

"Well, during the day the land gets hot and the water stays cool, and during the night the land gets cool and the water is warm."

我们是住在岛上的，对吧？

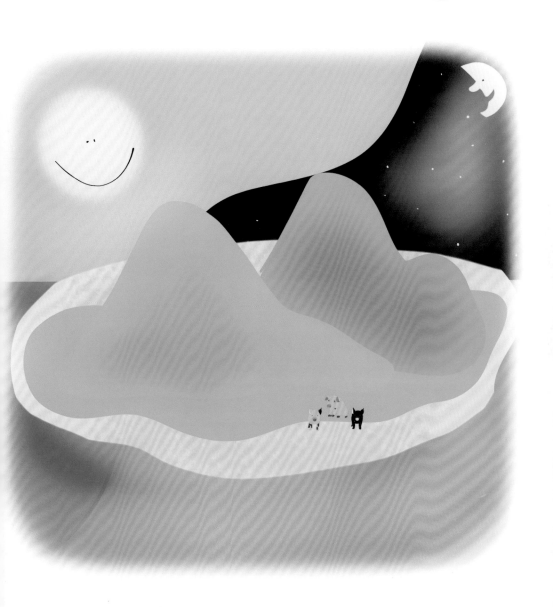

We live on an island, right?

冷藏，然后用船把胡萝卜蛋糕运到世界各地

Freeze and ship carrot cake around the world

"这和胡萝卜蛋糕有什么关系吗？"

"这意味着我们全年都有风啊。这样我们就有能量来切胡萝卜，利用免费的能源烘焙、冷藏，然后用船把胡萝卜蛋糕运到世界各地。"

"别忘了，在地球的另一端，胡萝卜蛋糕还得解冻呢。"

"噢，那样他们就可以卖新鲜出炉的美味蛋糕了。"小猪微笑着说。

"What does that have to do with carrot cake?"

"It means that we have wind all year. We have the power to shred carrots, bake, freeze and ship carrot cake around the world with free energy."

"Don't forget that on the other side of the world, the carrot cake will have to be unfrozen."

"Ah, but then they can sell great cake oven-fresh," smiles the pig.

"这些销售员总能找到好听的词。你怎么能把已经冷冻了几周甚至几个月的东西叫做'新鲜出炉'的呢?"

"不要担心这些用词。我们的胡萝卜蛋糕能实现让我们的风出口。"小猪说。

"我希望你是指空气的风,而不是我们肚子里的风!"他的朋友大笑着说。

……这仅仅是开始!……

"These sales people find nice words for everything. How can you call something that's been frozen for weeks and months 'oven-fresh'?"

"Don't worry about words. Our carrot cake will allow us to export our wind," says the pig.

"I hope you mean wind from the air, not wind from our tummies!" laughs his friend.

... AND IT HAS ONLY JUST BEGUN!...

……这仅仅是开始！……

...AND IT HAS ONLY JUST BEGUN!...

Did You Know?
你知道吗？

Supermarkets don't buy fruits and vegetables unless they are a prescribed shape, even if they taste great and the price is right.

超市进货时不会选择那些外形不规则的水果和蔬菜，就算味道鲜美、价格公道也不行。

Newborn piglets will leave the nest to go to the ablution area within hours of birth. Pigs can by potty-trained like dogs – as a matter of fact, pigs are easier to train than dogs.

小猪在出生后几小时内就会离开猪窝，被送去冲洗干净。猪可以像狗那样被训练如厕，事实上，猪比狗训练起来更容易。

Pigs have 4 downward-pointing toes on each foot, so a pig walks on the tips of its toes rather than on its whole foot.

猪的每只蹄子上都有4个趾尖向下的脚趾头，因此，猪是靠脚尖而不是整个脚掌走路。

Pigs need to chew their food, otherwise it cannot be digested. Pigs have a tremendous sense of smell.

猪必须咀嚼食物，否则不能消化。猪的嗅觉非常灵敏。

Pigs constantly communicate with each other. They have a range of different oinks, grunts and squeals, each with its own meaning.

猪时常会和同伴交流。它们会发出哼哼声、咕噜声和尖叫声等不同的声音，每一种声音都代表着不同的含义。

The pig is the last of the 12 animals in the Chinese zodiac. It represents fortune, honesty, happiness and virility.

在中国的十二生肖中，猪排在最后一个。其寓意是财富、诚实、幸福和刚强有力。

Coastal zones always have wind: it is created by a difference in temperature and pressure between the land, which absorbs more heat during the day, and the cooler sea. The result is a permanent flow of wind.

海岸地带总是有风。这个风是由陆地与海洋之间的温差和气压差产生的，白天，陆地吸收热量较多，海洋则相对凉爽，结果产生了持久的气流。

The term "oven-fresh" actually means recently unfrozen and thus does not at all refer to fresh food.

"新鲜出炉"这个词实际指的是刚刚解冻，根本不是指新鲜食物。

Think About It
想一想

Do you like eating fruits and vegetables because they look good, or because they taste good?

你喜欢吃水果和蔬菜，是因为它们外形好看，还是因为它们味道鲜美？

如果一个地方总是刮风，却没有将风力利用起来，你认为是什么原因？

If it's always windy somewhere, what reason is there for not harvesting wind power?

Do you think that bread called "oven-fresh" is really fresh?

你认为号称"新鲜出炉"的面包真的新鲜吗？

我们应该把不要的胡萝卜给猪吃，还是应该先榨取胡萝卜汁然后再把剩下的胡萝卜渣喂给猪吃呢？

Should we feed left-over carrots to pigs, or should we first get the juice out and then feed the pigs the waste?

Do It Yourself
自己动手

Go to the supermarket and look for carrots. Are they all the same shape? Then go to a farmers' market and look for carrots. Do they look the same as the ones at the supermarket? Then go back to the supermarket and ask some customers if they would mind tasty but odd-shaped carrots. Interview at least 10 people, starting with your mom and dad. Record their reaction when you ask them if they like "odd-shaped" or "ugly-looking" carrots. The interview will be more interesting if you have a few unusually shaped carrots at hand.

去超市里观察一下胡萝卜。它们都是相同形状的吗？再到农贸市场去观察一下，这里的胡萝卜和超市里的一样吗？再回到超市找几位顾客，问一下他们是否介意购买那些味道好但形状古怪的胡萝卜。至少要采访10个人，可以先从自己的爸爸妈妈开始。当你问他们是否喜欢"奇形怪状"或"长相丑"的胡萝卜时，记录下他们的反应。如果你手里拿着几个奇形怪状的胡萝卜，采访会变得更加有趣。

TEACHER AND PARENT GUIDE

学科知识
Academic Knowledge

生物学	一个拥有5000头猪的养殖场所产生的粪便量与一座拥有5万人口的城市所产生的相当；在欧洲，法律禁止用肉或骨粉饲料来喂猪；在猪作为禽流感病毒的中间宿主时，该病毒可以感染人类。
化　学	吃三根胡萝卜就能提供足够走完三英里的能量；胡萝卜中的β-胡萝卜素让猪肉呈现出更饱满更深的颜色。
物　理	胡萝卜呈橘黄色是因为它吸收了特定波长的光线，尤其是蓝色和靛蓝色；猪为了凉快而在泥里洗澡；科里奥利效应是指在旋转的坐标系统（旋转的地球）中运动，物体的路径会发生偏移。
工程学	胡萝卜收割机被设计成连刚长出的嫩胡萝卜都可以收割。
经济学	中国和俄罗斯的胡萝卜产量分别名列世界第一和第二；在鸡尾酒派对上提供小胡萝卜作为点心是一种营销策略，通常超市不卖这种小胡萝卜，这种做法盈利丰厚，以至于一些公司开始购买大胡萝卜然后切成小块出售；经商的秘诀，在于要有库存（仓储）和物流（逐日供应）。
伦理学	我们怎么能拒绝出售那些完全能吃，只是长得不好看的水果和蔬菜呢？
历　史	猪在11000年前被驯化家养；胡萝卜的种植起源于中亚和中东地区，最初的颜色有红色、紫色和黄色；胡萝卜蛋糕从中世纪的胡萝卜布丁发展而来，因其天然甜美的味道广受称赞。
地　理	作为向王室进贡的贡品，橙色的胡萝卜在荷兰非常流行，还有一个原因是橙色胡萝卜比其他颜色的胡萝卜更甜，肉质更饱满。
数　学	圆锥形的几何结构是什么样子？
生活方式	全世界大约有20亿头猪，而且猪越来越被人们接受作为宠物；猪的每个部位都能吃；德国泡菜（酸卷心菜）是由包括明串珠菌和乳酸杆菌的乳酸菌发酵而成的，通常配着猪肉或猪肉香肠吃。
社会学	胡萝卜最早是当作药物来种植，而不是食物；关于猪的习语有很多，但是在西方世界，这些习语基本上没有正面含义，而且也经常是错误的："吃得像猪似的"——其实猪吃东西很慢而且咀嚼地非常仔细，"像猪似的全身冒汗"——其实猪没有汗腺，根本不会出汗。
心理学	画猪人格测试，是指你在纸上画一只猪，这是一种让人们发现自己隐性人格的有趣方式，但是在人格评估中是不准确的；习语"胡萝卜加大棒"是指提供物质奖励（胡萝卜）辅以惩罚（大棒）的威胁来激励某人去做事，来源于马车夫在骡子面前悬挂一根胡萝卜来驱使其往前走。
系统论	那些因长得"丑"而被扔掉的胡萝卜除了为制作胡萝卜蛋糕提供丰富的原料外，还为猪提供了高质量的饲料。

教师与家长指南

情感智慧
Emotional Intelligence

猪

第一只小猪向同伴抱怨时非常有自信，提醒后者咀嚼完切碎的胡萝卜也是非常耗费时间的。他很好奇为什么有这么多美味的胡萝卜可以吃，而且也想知道为什么超市不喜欢这些胡萝卜。小猪意识到吃胡萝卜的好处：他的肉会变得粉嫩，很讨消费者喜欢。小猪们发起了一段积极的对话，而且好奇心很强。他们选择寻求让各方都受益的解决方案：省下胡萝卜汁，然后将胡萝卜残渣作为饲料。在探寻过程中，小猪们非常现实，考虑到了方案的缺点，比如为保证一年的供给而需要储存胡萝卜和冷藏胡萝卜所需的能源成本。下定决心的小猪们观察到岛上常年刮风的现象，于是想出了利用风力来满足加工胡萝卜所需的额外能源的解决办法。小猪们每想出一个方案，自信就会增加一些，他们接受了有创意的营销策略，并开始设想独到的销售计划——用风力出口胡萝卜。小猪们很知足，十分自在地开着玩笑。

艺术
The Arts

猪呼噜噜的叫声在不同的语言里有不同的表达。你想知道都有哪些表达吗？实际上，猪能发出20种发音。在以下这些语言中，猪叫声听起来像什么呢？试着来些和声吧。

中文：呼噜呼噜
丹麦语：øf
英语：oink oink
芬兰语：röh röh
法语：groin groin
匈牙利语：röf-röf-röf
日语：buubuu

韩语：kkool-kkool
挪威语：nøff-nøf
波兰语：chrum chrum
俄语：khryu-khryu
西班牙语：oink-oink
瑞典语：nöff

TEACHER AND PARENT GUIDE

思维拓展
Systems: Making the Connections

对于水果和蔬菜的形状和外表，消费者心中已经有了固定的想法。可能有多达25%的水果和蔬菜会被视为长相稀奇古怪，因而不被超市购买。这对农民来说是非常大的损失，评级高的胡萝卜售价可高达每吨1000欧元，而作为动物饲料的胡萝卜售价只有每吨15欧元。如果超市和消费者在喜欢味道和质量的同时，也能够接受那些小疙瘩和瑕疵，那么同样的收成，农民可以多赚20%。但这不意味着养殖场的农民要去买高价饲料，而是说我们要在获取食物和营养方面变得更明智些，在能源利用方面变得更高效些。果汁产品是价值链中的重要组成部分，可以创造更多的就业机会：胡萝卜汁是价值较高的饮品，而剩下的胡萝卜渣通常却被扔掉了，虽然它们对于动物（尤其是猪）来说很容易消化。把奇形怪状的胡萝卜做成胡萝卜蛋糕和腌胡萝卜，为消费者提供了价值更高的产品。尽管加工过程需要更多能源，但是我们却常常忽略对本地可再生能源的充分利用，也就失去了特有的竞争优势。瑞典哥特兰岛生产的本地胡萝卜蛋糕就受益于这个策略，即将价值较低的原材料转化成备受欢迎的产品——胡萝卜蛋糕，烘烤后出口到全世界。收割、储存、加工、烘焙、冷冻、包装、运输、解冻等工序所需的能源都是由岛上的风车所提供的，因此创造了远超过想象的收入。风电在该国的售价和电网电价一致，使得风力发电的投资在不到5年的时间内便可收回。这种以当地资源为主导的生产和消费一体化设计，可以在农民、公司和社区之间创造更好的灵活性。

动手能力
Capacity to Implement

胡萝卜蛋糕很容易做，腌胡萝卜就更简单了！将干净的胡萝卜放进罐子里，将罐子填满到再也塞不进胡萝卜为止。将浓盐水倒在胡萝卜上，直到全部没过，然后盖上松紧合适的盖子。常温放置24小时，时不时检查一下是否有气泡，这些气泡闻着有新鲜的酸味。当出现气泡后，把罐子放进冰箱，再过一两个星期就可以吃了。如果一次吃不完，就把剩下的放进冰箱里最凉快的地方。为什么要吃发酵了的胡萝卜呢？提出论点，然后给出一个逻辑性强、有说服力的答案。同时，你还可以开启自己的腌萝卜事业，将那些奇形怪状的胡萝卜做成美味健康的食物。

教师与家长指南

故事灵感来自

哈肯·埃尔斯腾
Håkan Ahlsten

哈肯·埃尔斯腾生来就注定会成为一名银行家。他出生于瑞典的哥特兰岛,一直承诺要为下一代将这座岛发展到最好。他受到所看到的机遇的启发,于是决定离开银行,成为一名企业家。他的创举之一便是接管了当地的一家小面包房,并将其发展成为瑞典胡萝卜蛋糕供应商的龙头企业。他让农民送来长相奇怪的胡萝卜,动员另一位企业家创建了胡萝卜储藏企业 Ryftes,以确保每日供应。出口海外的胡萝卜的价格里包含了收集风力所需的成本。现在,哈肯和他的妻子玛利亚正在哥特兰岛上经营一家出色的精品酒店。

更多资讯

http://www.think-differently-about-sheep.com/Sentience-%20In-Farm-Animals-%20Pigs.htm

www.telegraph.co.uk/news/worldnews/europe/france/11131994/Ugly-fruit-and-vegetables-prove-a-hit-in-France.html

http://www.gragasen.se/

http://weather.mailasail.com/Franks-Weather/Coastal-Wind-Variatios

图书在版编目（CIP）数据

风吹来的蛋糕：汉英对照 /（比）鲍利著；王菁菁译． -- 上海：学林出版社，2015.6
（冈特生态童书．第2辑）
ISBN 978-7-5486-0863-9

Ⅰ．①风… Ⅱ．①鲍… ②王… Ⅲ．①生态环境－环境保护－儿童读物－汉、英 Ⅳ．① X171.1-49

中国版本图书馆 CIP 数据核字（2015）第 086058 号
——————————————————————————————

© 2015 Gunter Pauli
著作权合同登记号 图字 09-2015-446 号

冈特生态童书
风吹来的蛋糕

作　　者——	冈特·鲍利
译　　者——	王菁菁
策　　划——	匡志强
责任编辑——	匡志强　蔡雾奇
装帧设计——	魏　来
出　　版——	上海世纪出版股份有限公司学林出版社
	地　址：上海钦州南路81号　电话／传真：021-64515005
	网　址：www.xuelinpress.com
发　　行——	上海世纪出版股份有限公司发行中心
	（上海福建中路193号　网址：www.ewen.co）
印　　刷——	上海图宇印刷有限公司
开　　本——	710×1020　1/16
印　　张——	2
字　　数——	5 万
版　　次——	2015年6月第1版
	2015年6月第1次印刷
书　　号——	ISBN 978-7-5486-0863-9/G·312
定　　价——	10.00 元

（如发生印刷、装订质量问题，读者可向工厂调换）